OIL

THE ECONOMICS
OF FUEL

Joann Jovinelly

ROSEN
PUBLISHING®

New York

Published in 2008 by The Rosen Publishing Group, Inc.
29 East 21st Street, New York, NY 10010

Library of Congress Cataloging-in-Publication Data

Jovinelly, Joann.
Oil: the economics of fuel / Joann Jovinelly.
   p. cm. — (In the news)
Includes bibliographical references and index.
ISBN-13: 978-1-4042-1915-1 (hardcover)
ISBN-10: 1-4042-1915-3 (hardcover)
1. Petroleum reserves. 2. Energy consumption. I. Title.
HD9560.5.J68 2008
333.8'232—dc22

2007008759

Manufactured in the United States of America

On the cover: *(Clockwise from top right)* A gas tank is filled with fuel; crude oil is pumped in 2001, at the Kern River Oil Field in California; and the SK Corporation oil refinery in Ulsan, South Korea, one of Asia's leading energy and petrochemical companies.

contents

1 The Oil Economy   **4**

2 The Golden Years of U.S. Oil   **18**

3 What Is Peak Oil?   **28**

4 Oil Consumption and the Environment   **33**

5 Beyond Oil:
   Alternative Energy Sources   **39**

Glossary   **55**

For More Information   **58**

For Further Reading   **60**

Bibliography   **61**

Index   **63**

The Oil Economy

Oil, which comes from petroleum, is a nonrenewable natural resource. Modern life as we know it is made possible largely because of the energy that has been harnessed by oil. In fact, most products and daily activities are made possible or made easier because of oil and its by-product, gasoline. Oil and gas help us cook our meals, heat our homes, and transport food, people, and products across the country and the world. Oil is the foundation for a variety of chemical products, including fertilizers, solvents, lubricants, and pesticides. Oil-based ingredients are also key in many items such as plastics, detergents, medicines, asphalt, and synthetic rubber.

The overall importance of oil cannot be overstated. The industrial revolutions of the last 200 years and all of the conveniences of modern life have been made possible because of man's ability to transform the fossil fuels oil, natural gas, and coal into energy.

Methane bubbles rise from the natural oil deposits known as the La Brea Tar Pits located in Los Angeles, California. Based on fossils found there, scientists estimate the pits are at least 40,000 years old.

What Is Petroleum?

Petroleum oil is an ancient substance that was created naturally over the course of millions of years. It formed from the remains of plants and animals that were compressed over time and fossilized (this is why petroleum is frequently referred to as a fossil fuel). Almost from the moment that petroleum comes into existence, it seeks the surface of the earth. As its trapped gas bubbles expand, the liquefied oil-gas mixture moves steadily outward or upward. As the oil-gas mixture rises, it separates because the gas is lighter than the oil. This oil is called crude oil.

A few areas of the world have known about crude oil for centuries. It surfaced naturally in the Middle East, including in places now known as Saudi Arabia, Iraq, and Iran. It even was used in early warfare. Greek fire techniques, employed as early as 670 CE, used devices attached to the prows of ships that shot continuous flames.

It would take centuries, however, for humans to understand the importance of crude oil and its potential to provide a long-lasting, versatile, and abundant source of energy. Early discoveries of crude oil in America during the nineteenth century transformed the United States into the most powerful and influential country in the world. This is primarily because the success or failure of any industrialized country is based largely upon its ability to harness a reliable and inexpensive energy source.

Oil Is a Commodity

Oil is a raw material, or commodity, that is bought and sold on the world market daily. When a person buys a commodity, he or she is purchasing a contract, not an actual product. This contract, sometimes called a futures option, is an agreement to buy or sell a product in the future at a specific price. Buying and selling commodities—whether in oil or gold or wheat or sugar— is a little like speculating that the value of the commodity will rise or fall based upon its global supply and demand. (If the price of the commodity rises on the agreed-upon date, for instance, the value of the contract increases, earning the owner a profit when he or she sells it.) The supply and demand for oil influences the United States and world economies more than any other commodity. Naturally, any increase or decrease in the supply or demand for oil (such as conflicts between nations, extreme weather conditions, or discoveries or depletions of petroleum at existing drilling sites) will also affect its supply, production, and distribution. Together, these changes ultimately influence the price at which oil is traded within the world's financial markets.

Supply and Demand

The price of oil, which is measured in barrels, is continually rising and falling based on its worldwide supply and demand. These fluctuations also affect the price of gasoline. Even when supplies of oil are stable, its price per barrel rises and falls. This is a normal response to factors such as the change of seasons (demand for oil increases during the winter months when heat is needed, and in the summer when many people drive

long distances on vacations) and ordinary competition between suppliers. Any increase or decrease in the price of oil also influences other businesses and industries because the cost to produce and distribute goods rises when the price of oil increases.

Another factor that can greatly influence oil prices is when some of the world's greatest suppliers of crude oil, such as nations like Saudi Arabia, decide to increase their oil production. This floods the market and ultimately sends the price per barrel tumbling. The oil-producing nations of the world therefore attempt to stabilize prices by working together to control its production. Over time, these controls help to guarantee steady, constant revenues.

An example of decreased supply and increased demand was seen in August 2005 when Hurricane Katrina hit the U.S. Gulf Coast. At least twenty refineries off the coast were heavily damaged or destroyed. Because it was feared that U.S. oil production would be cut by at least one-third, prices immediately soared. In the months following the hurricane, they reached roughly $70 per barrel, an increase of about $10 per barrel. According to the Energy Information Administration (EIA), Hurricane Katrina "took out more than 25 percent

Oil rigs pump near Los Padres National Forest, California. Under a plan supported by the Bush administration, formerly protected U.S. areas potentially rich in oil may soon be opened to commercial exploitation.

This oil platform was ripped away from its mooring in the Gulf of Mexico by Hurricane Katrina in 2005. Natural disasters can increase oil prices by damaging facilities and therefore decreasing production.

of U.S. crude oil production and 10 to 15 percent of its refinery capacity." The large drop in supply ultimately increased oil and gas prices. After Katrina, some gas stations in the United States were charging approximately $6 per gallon for gasoline.

Geopolitics

Geopolitics refers to the politics and policies of the world's nations, their geographical locations and natural resources, and how they relate to one another. The primary influences over the world's oil include the United States, since it was the first to lead in oil production and is the country with the greatest demand. The countries of the Middle East, particularly Saudi Arabia, are also important because of their vast petroleum assets and reserves. Lastly, oil's price, which is always changing, plays a significant role in the political negotiations regarding the buying and selling of fossil fuels.

The United States has had great influence over the oil market throughout its history. Not only was America once the leader in world oil production, it remains the third-largest oil producer today. In addition, America consumes one-fourth of all the oil produced in the world. The country also had the fastest-growing consumption rate during the 1990s, increasing roughly 3.5 billion barrels a year throughout the decade. Saudi oil exports to America were roughly 1.5 million barrels per day from January to May 2005, or about 15 percent of the foreign oil imported.

Saudi oil companies sell oil to American companies at a discount to protect the steady revenue that helps make them increasingly wealthy. When it comes to profits, most countries turn a blind eye to disagreements or cultural differences. For instance, the United States imports oil from a variety of oil states with anti-American attitudes, including Syria, Libya, and Venezuela.

Just as in any other discussion about economics, supply also has a distinct role in geopolitics. The fact is that no one person clearly knows how much oil is available in Saudi Arabia or in the Middle East in general. Experts estimate that Saudi Arabia alone possesses about a quarter of the world's remaining reserves, clearly enough for the country's leaders to have an ongoing influence in geopolitics. In addition, the Saudis' oil is also easy to access. They therefore could overpower the crude market at any point by offering an abundance of less expensive oil for sale. This thus would lower oil's price per barrel on the world market and eliminate competitors.

On the positive side, Saudi exports help stabilize the market when conflicts occur, such as wars between nations, market turbulence, or when oil drilling in exporting countries goes offline or is stopped due to bad weather. For instance, when the United States could not import oil from Iraq during the First Gulf War, Saudi oil companies picked up the slack in the market by increasing their exports.

The changing price of oil is the final major influence on its global market. Oil's price per barrel determines how much money oil companies and the people who

OPEC president Edmund Daukoru *(center)* and secretary general Mohammed Barkindo *(right)* welcome Venezuelan oil minister Rafael Ramirez to Nigeria in December 2006. They met to discuss plans for reducing oil production within OPEC member countries in order to stabilize global oil prices.

invest in oil earn. Price also influences how much oil is bought, so it also dictates oil's availability.

OPEC

The Organization of the Petroleum Exporting Countries (OPEC) is an intergovernmental agency that was created in 1960 by Iran, Iraq, Kuwait, Saudi Arabia, and Venezuela. (Over time, other oil-exporting countries joined, including

Qatar, Indonesia, the United Arab Emirates, Algeria, Nigeria, and Ecuador, among others.) Today, OPEC member nations control more than 50 percent of the world's oil supply.

OPEC's main goal is to stabilize oil prices throughout its member countries. This avoids significant ups and downs in the market, which can be disruptive and damaging to the global economy. Of course, individual oil companies are bound to make attempts to manipulate the market by purposefully selling less oil, thereby increasing demand to drive up prices. Or they flood the market with cheap oil, which drives global prices down. Generally, economists consider these manipulations to be very destructive. U.S. leaders and oil company executives routinely pressure OPEC to help regulate oil prices when they get too high or too low. This is usually done by increasing or decreasing production, which, in turn, increases or decreases supply.

In many instances, such as during the First Gulf War and after the 9/11 attacks on the United States, OPEC nations have provided America with increased oil imports. This helps keep the global supply and price of oil steady, preserving the global economy. OPEC has had its ups and downs over the years, but it is generally viewed as a valuable organization that is necessary to help control a volatile oil market.

Oil and the War in Iraq

It is possible that the U.S. war in and occupation of Iraq has had more to do with having influence over Iraq's vast oil resources than eliminating weapons of mass destruction or bringing democracy to a dictatorial land. One supporting opinion is from journalist Greg Palast, who told BBC News, "The Bush Administration was involved in making plans for war and for Iraq's oil before the 9/11 attacks." Palast reported that the United States planned to sell off Iraq's oil in order to destroy OPEC's control over the price per barrel by increasing production and flooding the world market with cheap oil.

Ultimately, the price of oil determines the flow of international money (countries will purchase the least expensive oil available) and the muscle of political influence. By keeping the price of oil low, the United States also is discouraging conservation and the development of alternative energy sources.

The Strategic Petroleum Reserve

The United States can influence price-per-barrel cost by releasing oil from its Strategic Petroleum Reserve (SPR). The SPR is a stockpile of government-owned crude oil. It can be shared with oil companies to sell to the American

One hundred and sixty million barrels of oil are stored in this location of the Federal Strategic Petroleum Reserve (SPR) facility near Beaumont, Texas. The SPR is meant to supply Americans with oil in case of severe disruption in oil imports.

people if the importing of foreign oil suddenly becomes disrupted or unavailable. This stockpile is also for national defense should foreign nations implement barriers to importing. The SPR was established in 1975 after just such a situation: the Arab oil embargo of 1973–1974. During this period, oil prices shot up because of the Yom Kippur War, a conflict between Israel and the combined Arab nations of Syria and Egypt. Oil prices also escalated in 1979 during the American hostage crisis in Iran. For

444 days, starting in November 1979, Iranian militants in Tehran held nearly seventy Americans hostage. In both of these events, Arab nations either limited or completely disrupted American access to foreign oil imports. Oil was also released from the SPR after Hurricane Katrina.

The SPR, which currently contains an estimated 700 million barrels of crude oil, is located in four sites off the Gulf Coast, some 2,000 feet (610 meters) below the surface of the earth. In 2007, President Bush decided to double the capacity of the SPR, a decision that was met with some surprise since it has rarely been used since 1975. The current plan is to invest $65 billion dollars to harvest more crude oil, not including the $2 billion dollars it costs per year to run and secure the facilities. Still, this is a fraction of what is spent by the government in direct subsidies to oil and gas companies in order to keep prices down and stimulate economic growth.

The Golden Years of U.S. Oil

THE STANDARD OIL C
AUTO FILLING STATI

A merica's early settlers spoke of a land that had abundant natural resources, and was rich in animal life and dense with thick forests. It was those forests that became a primary energy source for hundreds of years. Homes were heated and food was prepared with wood-burning stoves. Wood was used to build windmills and water mills, just as it had been in medieval Europe. Wood also was used to make charcoal, which could be heated to degrees high enough to smelt iron and make tools.

Coal

Before long, wood supplies dwindled, and Americans turned to coal for their energy needs. Although coal had been used for centuries (it was used in China for more than 4,000 years), it was regarded as a poor source of energy because of the black soot that it produced.

Photographer Lewis Wickes Hine took this image of two employees of the Pennsylvania Coal Company in an underground coal mine in Pittston, Pennsylvania, in 1911. Hine helped increase public awareness about the dangers that miners faced.

For instance, coal was the primary energy source in thirteenth-century England, and the city of London was blackened by dark smoke during the winter months.

In America, as more people realized that wood was in short supply, coal became a useful alternative. It could be heated to higher temperatures than wood or charcoal, producing a hot-burning fuel called coke. The coke was then used to forge both iron and steel. By the seventeenth century, coal was being used to produce a variety of other products, too, including glass, tile, and brick.

Still, as useful as coal was, it was difficult to extract. And coal mining was dangerous. Many miners were injured or killed during underground explosions. But coal mining also led to important inventions, such as the elevator, secure underground tunnels, and by the 1820s, the first steam railroads. Soon a variety of other steam-powered inventions were produced, and the economies of England, Europe, and Russia were booming. Artificial gas, the manufactured by-product of coal, illuminated the streets of London in 1807 and Baltimore, Maryland, in 1816. The Industrial Revolution was in full swing. Huge factories were able to mass-produce machine-made goods, and coal-powered cargo steamers shipped those goods across the world. In America, coal was the main source of energy into the twentieth century. (America still has considerable coal resources; experts believe coal and nuclear power are going to satisfy U.S. energy needs as oil grows increasingly scarce.)

A Petroleum Renaissance

The United States was once the world's leader in oil production. The first major discovery of petroleum on its soil dates back to 1859. Edwin L. Drake, a retired railway conductor, powered the world's first successful oil drill near Titusville, Pennsylvania, with an old steam engine. Soon dozens of drillers and refiners were expanding the supply of petroleum. At first, it was used as a machine lubricant. Kerosene (a petroleum by-product) was used as lamp fuel.

Within the decade, the first oil entrepreneur, John D. Rockefeller, was involved in the move toward an oil economy. By 1866, he had begun purchasing and refining massive amounts of oil. His new company, Standard Oil, became the largest oil refining operation in the country. Within a few years, he was buying up his competitors, quickly making Standard Oil the number-one energy company in America. Rockefeller undercut his competitors, thereby driving them out of business. He made up the difference by getting kickbacks from the railroad companies that transported his (and other companies') oil. Standard Oil executives blocked new businesses from opening and threatened or blackmailed competitors. Within a short time, Rockefeller had enough money to further develop his empire, building his own pipelines and oil tankers. Standard Oil, now worth millions, had become a monopoly. By 1880, it was America's number-one exporter and in control of 90 percent of the country's oil business.

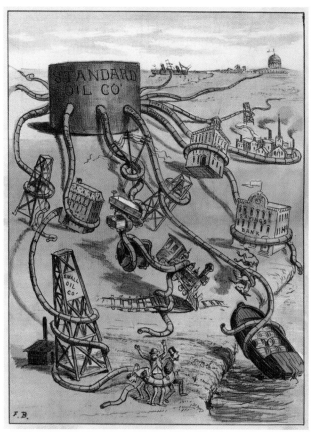

This 1884 political cartoon *(left)* depicts the Standard Oil Company monopoly as an octopus, crushing other businesses and industries. John D. Rockefeller *(right)*, president of Standard Oil, was a visionary entrepreneur who immediately understood oil's importance.

Early Competition in the Oil Market

In the meantime, oil was discovered in Russia in 1871. Like Rockefeller, Ludwig Nobel (brother of Alfred Nobel, after whom the Nobel Prize is named) quickly established an oil business. Within a few years, it had reached a third of America's oil production. With financial help from the Rothschild banking family, Nobel began building a series of railroads throughout Russia and in the Caspian Sea region. The railroads transferred oil to European markets, territory formally served by America's Standard Oil. (In response, Standard Oil established an affiliate in England called the Anglo-American Oil Company.) Within time, the Rothschilds expanded their oil operations in Russia, exporting first to Britain and then to Asia. There, they joined forces with Marcus Samuel to build a system of transport. Eventually, Samuel was transporting oil around the world. By the 1890s, the Oil Wars between Rockefeller, Nobel, the Rothschilds, and Samuel were under way. Oil also had been discovered in the Dutch East Indies (present-day Indonesia), further eliminating Rockefeller's grip as the world's petroleum giant.

Electric light was invented thanks to Thomas Edison and Nikola Tesla, and the use of kerosene started to wane. Oil was beginning to be used for home heating and to power ships, trains, and factories.

Early in the 1900s, oil was discovered in Persia (present-day Iran), which led to the establishment of the Anglo-Persian Oil Company (now BP, or Beyond Petroleum). In 1911, Standard Oil became Standard Oil of New Jersey and was broken up into several smaller businesses under the Sherman Antitrust Act. It then became Standard Oil of New Jersey (Exxon), Standard Oil of California (Chevron), Standard Oil of New York (Mobil), Standard Oil of Indiana (Amoco, now BP), Continental (Conoco), and Atlantic (now BP). Increasing oil discoveries in Texas, California, and Oklahoma were taking shape (as were the companies Texaco, Gulf Oil, and Shell), driving production to greater heights. It was these new discoveries that helped America become the greatest foreign oil exporter of the first half of the twentieth century. By mid-century, seven American-owned companies (Exxon, Chevron, Mobil, Gulf, Texaco, Amoco, and Shell) controlled the majority of the oil reserves outside the United States and Russia, two-thirds of all oil tankers, and the majority of the world's oil pipelines.

U.S. Demand Outpaces Production

By the 1930s, American motorists were taking to the country's roads in great numbers and gasoline was becoming the main by-product of oil. Aviation fuel was also beginning to account for a noticeable share of oil use. Soon, an entire chemical industry would develop from petroleum, launching products from nylon to a variety of plastics. In the 1950s, petrochemical-based pesticides

American carmakers' earnings skyrocketed during the 1930s. According to a 1936 article in *Time*, from 1934 to 1935, earnings for General Motors increased from almost $94.8 million to $167 million.

were introduced, increasing food production and yields. Cities started demanding better roadways, people wanted their own cars, and the U.S. automobile economy was blossoming. The demand for private vehicles and for highways also helped secure the success of other major industries, including steel, rubber, plastics, and glass. Starting in the 1950s, passenger jet service was the next huge industry to be revolutionized by abundant oil availability.

At the end of World War II (1945), the United States was the world's largest consumer of oil. However, America could no longer keep pace with its own demand. At the same time, countries in the Middle East emerged as the largest oil producers. As their supply of oil exceeded demand, Middle Eastern nations formed OPEC to keep competition at bay and help keep prices from dropping. In 1959, oil was discovered in Libya. Because of Libya's proximity to European markets, a glut of available oil threatened to disrupt business and lower prices, further underlining the need for OPEC.

By the 1970s, American oil production had waned, and the United States began to have less input in stabilizing global oil prices. This became especially apparent as conflicts erupted between Israel, a nation allied with the United States, and Arab countries such as Egypt, Libya, and Syria. The U.S. relationship with Israel has, at times, made it more difficult to import oil from OPEC member nations. During times of conflict, such as during the 1973–1974 Yom Kippur War, for instance, the United States' relationship with Israel led to an Arab oil embargo, cutting off America from Middle Eastern oil imports. This artificial disruption in oil supply set off the 1970s' "gas crisis" in the United States. Higher prices resulted, as did long lines at gas stations. Domestic inflation, coupled with an unstable Middle East, led to continued price

Above, California motorists wait in lines at a gas station during the 1979 gas crisis. Because of the fuel shortage, the quantity of motorists' purchases was also restricted.

hikes and disruptions in oil imported throughout the 1980s. The United States started investing in nuclear power to offset some of its oil use. Meanwhile, new oil discoveries in the North Sea, Angola, Mexico, and off the coast of Nigeria also boosted crude oil production around the world, offering America access to other importing sources.

What Is Peak Oil?

More and more, scientists, economists, and politicians are using the term "peak oil" to describe what is thought to be the high point of available petroleum resources. The term, which has been in use for more than fifty years, means that at a specific point in time, humankind will reach the peak of the planet's extractable oil supply. From that point, the remaining available oil will be in a state of steady decline.

When experts say that we have reached peak oil, they mean that we have used the easiest-to-access half of the underground oil supply. Because the other half is so difficult to extract, and because it will take more and more energy to extract it, we will never be able to drill for all of it. Scientists use the phrase "energy return on energy invested," or EROEI, to indicate how much energy it takes to access the world's various fossil fuels. (If the EROEI is too low, a company would stop drilling for oil or mining for coal because it would not be profitable to do so.)

The peak oil theory, first put forth by the American geophysicist Marion King Hubbert, is sometimes referred to as Hubbert peak theory. When Hubbert presented his theory to the American Petroleum Institute in 1956, he predicted that the United States would reach the climax of its own oil resources between 1965 and 1970. His research was incredibly accurate; it turned out that the U.S. peak of oil resources occurred in 1971. Hubbert further concluded that the world's oil availability would peak about fifty years after he first presented his theory, or around 2006.

While working toward his doctorate at New York's Columbia University in the 1930s, M. King Hubbert (above) taught classes in geophysics.

When Will Peak Oil Occur?

Exactly when the world's peak oil occurs cannot accurately be known until years from when resources begin to dwindle. Like Hubbert predicted, the world may have already passed the point of peak oil, which is among the reasons why economists believe that oil prices have risen and will continue to rise. However, others with knowledge of the world's oil supplies challenge Hubbert's prediction. They say that oil discoveries in the North Sea (1960s) and Angola (1990s) ultimately counteract his research. In fact, it is simply not known how much undiscovered oil is left in the ground, so no one can exactly determine if peak oil has occurred or when it will occur.

Estimated years for when peak oil will occur are widespread and range from the present to 2035, based on proven oil discoveries. Many institutions, including the Association for the Study of Peak Oil (ASPO), along with representatives from Princeton University, the University of Colorado, Boston University, and others have all concurred that we have either already passed the point of peak oil, or we will do so soon. Others employed in the world's major oil companies are more optimistic. Their jobs and livelihoods are based around a steady supply of oil, and they maintain interest in investment and continuity. According to oil geologist Colin Campbell, who was interviewed at Global Vision Rio, an international peak oil conference in 2002, major oil companies like Exxon and BP are "very obtuse about what they say about [oil depletion]. The one [term] they don't like to talk about is oil depletion."

Paul Roberts, author of *The End of Oil*, said in a 2004 interview, "The world's oil supply will peak in about twenty-five years." But he also adds that it is very likely that most, if not all, of the rest of the world's production will peak before OPEC production peaks. Roberts believes that the non-OPEC production will peak within the decade. He is concerned about diplomacy when the world's most valuable natural resources are isolated in the Middle East. He also wonders how eager the Saudis will be to supply nations of the world with oil.

Comparative Oil Use Around the World

Global demand for oil continues to grow, especially in the United States, China, and India. The Energy Information Administration (EIA) predicts that this global demand will increase "60 percent by the year 2020 to roughly 40 billion barrels per year, or 120 million barrels per day." Oil reserves will not last forever. According to a 2006 article in *Harvard Magazine* by Jonathan Shaw, "Known global reserves of oil (based on current, not future, consumption), will last 41 years; natural gas 67 years; and coal 164 years."

Oil use in the United States outpaces that of all other nations and shows no signs of stopping. President Jimmy Carter put conservation programs in place in the 1970s as a response to high oil prices and long lines at the pump. President Ronald Reagan ended these programs in the 1980s. High fuel prices may have forced Americans to cut back on their overall energy consumption in the 1970s, but oil discoveries throughout the 1990s kept gas

Although the United States uses more oil than any other country, India and China are increasing their use of fossil fuels. Emerging cities such as Barakar, India *(above)*, for example, are accommodating more cars than ever before.

prices low and encouraged more driving. This period became the decade of the sports utility vehicle (SUV), which burns more gas than other cars.

Today, President Bush calls America's oil use an "addiction." The United States currently uses one-quarter, or 25 percent, of the world's oil. This amounts to approximately twenty million barrels of oil every single day. More than half of it is imported from other countries. The budget for America's oil use amounts to roughly $720 million each day.

Oil Consumption and the Environment

Most will agree that humankind is responsible for global warming and that the continual heating of our atmosphere is the direct result of releasing greenhouse gases into the environment. Unfortunately, all the years of burning inexpensive and readily available fossil fuels, like coal, oil, and natural gas, have contributed to our current warming condition. Steady increases of carbon dioxide emissions across the planet are trapping greenhouse gases in the atmosphere and storing heat closer to the surface of the earth.

In its 2007 report, the Intergovernmental Panel on Climate Change (IPCC) highlighted the human contribution to global warming. Its report concluded that the earth will soon be experiencing "increases in global average air and ocean temperatures, widespread melting of snow and ice, and rising global mean sea level." The panel's predictions range from temperature increases of "3.2 to 7.1 degrees Fahrenheit [1.8 to 4 degrees Celsius] by the year 2100." The threat of all of this constant warming

Homes in the Alaskan village of Shishmaref are being destroyed by beach erosion. These conditions have intensified due to global warming. Hundreds of families have already been displaced.

could, in fact, create an ice-free Arctic in the twenty-second century, and with it, the rise of sea levels by thirteen to twenty feet (4 to 6 m) around the world. Such a dramatic change is already in the works, along with more extreme weather patterns from an increasingly unstable environment. Scientists are predicting that over time, millions of people will be forced from their homes to become environmental refugees. At least hundreds of plants and animals species will face extinction, including polar bears and penguins. Risks to the water and food supply—due to increased flooding, desalinization, and drought conditions—are no less alarming.

Still, leaders in the Bush administration have only begun to acknowledge the strong warnings issued by the IPCC. In 2005, President Bush kept his word that he would not sign the Kyoto Protocol, an agreement between 141 nations that calls for a decrease in greenhouse gas emissions by 5 percent over the next decade. President Bush also did not make any significant policy changes in a response to the 2007 IPCC report. In fact, recent information has led some to believe that members of the Bush administration have purposely interfered with the release of information that they thought would have a damaging impact on oil company profits. According to U.S. representative of California Henry Waxman, who is chairman of the House Committee on Oversight and Government Reform, the committee has uncovered evidence that shows the Bush administration sought to "weaken government reports [about global warming] by emphasizing its 'beneficial effects' and downplaying its effects on human health."

In addition to the obvious need for changes to America's energy policy, world leaders also criticized Americans for their unabated energy usage, which today remains the highest of all the world's countries. (The rate of energy usage is expected to increase even more dramatically in China during the next decade.)

Reducing Carbon Emissions

Given the IPCC predictions, other countries around the world are trying to find solutions to reduce carbon emissions. They are seeking more clean-burning technologies as well as renewable energy sources like wind power, solar power, and hydropower. In the United States, as well as in other nations, there is a growing discussion about increasing the fines for pollution in order to decrease it. One method to accomplish this goal is to charge an energy tax for the emission of carbon dioxide into the atmosphere, an idea for which there is growing bipartisan support. Others are behind the idea of increasing mileage standards for all vehicles (so they would be more fuel-efficient). Yet another idea is carbon sequestration, purposely placing carbon underground or underwater so it cannot be released back into the environment where it will trap heat close to the earth. Most people who are concerned about reducing carbon emissions believe that there is not nearly enough discussion or support for basic energy conservation.

Energy Conservation

Because the earth can no longer sustain the increasing temperatures brought on by the constant burning of fossil fuels, energy conservation is important. The United

States has made great strides over the years in improving energy efficiency rates (using 47 percent less energy per dollar than thirty years ago). However, every individual can help to conserve energy. One of the ways in which the government can help to accomplish this is by raising fuel economy standards. This is likely to happen in the near future. There are plenty of things that you can do, too, such as:

This gas pump in Hudson, Wisconsin, offers E85 fuel (85 percent ethanol and 15 percent gasoline). America's ethanol is produced from corn crops.

- Use compact, long-lasting fluorescent light bulbs
- Unplug power strips (surge protectors) when not at home
- Buy energy-efficient ("Energy Star") appliances
- Use recycled paper products
- Cancel print catalogs
- Stop drinking individual plastic bottles of water
- Don't pre-rinse dishes before they are put in the dishwasher (and if you wash by hand, use a basin)
- Use fewer lights and turn them off when you leave a room
- Start a carpool

- Use mass transit, ride your bike, or walk to your destination
- Combine your errands into one trip per week
- Reuse bags and carry fabric bags to the grocery store
- Wash clothing in cold water
- Use glasses and mugs instead of disposable cups
- Line-dry clothing instead of using the dryer
- Open drapes to let sun in during the winter and close them to keep it out during the summer
- Keep tires properly inflated
- Wear a sweater instead of turning up the heat
- Use a fuel-efficient vehicle
- Turn off the car ignition while waiting to pick someone up
- Insulate your home's water heater

There are hundreds of ways to save energy and reduce carbon emissions. Start today.

Beyond Oil: Alternative Energy Sources

Americans generate roughly 26.5 tons (24 metric tons) of greenhouse gases per person per year. Most of these gases are released by burning gasoline. In President Bush's 2007 State of the Union address, he cited America's dependence on foreign oil imports and asked Congress to support a policy he called the Advanced Energy Initiative. In this plan, the United States, by creating new technologies and fuels, hopes to replace at least 75 percent of its oil imports from the Middle East by 2025. In several key areas, this initiative calls for research into clean-burning and renewable energy sources such as those derived from the sun and wind. It also calls for improvements to current U.S. technologies in order to reduce the high levels of pollution presently emitted by burning coal. The 2007 U.S. budget includes more than $280 million for this purpose. Another $50 million is on tap to encourage private companies to develop cleaner coal-burning technologies.

The Era of Declining Resources

More and more, economists and scientists assert that humankind is currently living in a period of declining resources. Most people now agree that even with new and increasingly powerful deepwater drilling technologies, and the potential to drill for oil in areas that were previously off-limits to American and foreign companies, we are living in an age of peak oil. Gone are the days when oil was so readily available that drilling and refining were extraordinarily profitable. As oil and natural gas become harder to find and more expensive to extract from the earth, the profits that can be made by selling them will decrease. In short, the era of "cheap" oil is in the past. As petroleum becomes more difficult to extract and its sources are sometimes less than pure (as in the rich tar sands of Alberta, Canada), more energy will be consumed to obtain the oil from the earth than can be had by refining it. This will result in a negative EROEI.

Consumption is another reason why many scientists agree that we are living in an era of declining oil resources. Even considering the potential for greater extractions in remote areas, the world's oil consumption (particularly in the United States and China) is growing too rapidly. This is mainly due to the world's increasing population. Some people thus believe that controlling population growth could be among the solutions to our growing energy needs.

Alternative Energy Sources

Besides conserving natural resources, increasing the use of alternative energies may also help forestall reaching the end of the world's oil supply. Successful replacements for oil must meet several important criteria. They must be easy to transport, provide a rich source of power, and be useful for a variety of different applications. The challenge so far is that few of the major alternative nonrenewable (natural gas and coal) and renewable sources (wind power, hydroelectric power, and solar power) meet all of these criteria.

Nonrenewable Energy Sources

Just as the world has a limited supply of oil, other non-renewable energy sources are also dwindling. These resources, mainly natural gas and coal, will likely be used with crude oil until they reach peak and become too costly to produce.

Natural Gas

According to the U.S. Energy Information Administration (EIA), natural gas currently accounts for approximately 25 percent of America's energy consumption. It is versatile, burns more cleanly than oil (although it still contributes to carbon dioxide emissions, and therefore to global warming), and its EROEI is high. However, natural gas, like oil, is a nonrenewable resource in decline. The demand for natural gas is also on the rise, especially as a source for electricity since Americans use more electrical appliances than ever before. The first outward sign of natural gas's decline occurred in 2000 when prices rose roughly 400 percent. Even with increasing natural gas imports from Canada, American demand is expected to overpower supply within fifty years, according to the U.S. Energy Information Administration. And even though there is an abundant supply of natural gas resources in the Middle East, it is not easy or cost-efficient to transport. All of America's natural gas supply comes from North American sources.

Coal

Coal, another nonrenewable source of energy, is also largely used for electricity. Although coal is among our most abundant resources (government sources claim the coal supply will last 200 more years), it emits a great deal of pollution when burned, including carbon dioxide and sulfur oxides. Although some coal companies speculate that coal is a useful and inexpensive replacement for oil, environmentalists disagree. In addition, companies generally use oil to mine coal, and its supplies are becoming more difficult to access. In Bill McKibben's *The End of Nature*, he discusses how the use of coal has been debilitating to America's soil and water by "weakening trees and acidifying lakes to the point of sterility."

The United States has committed more than two billion dollars since 2000 to developing cleaner coal-burning technologies. However, other countries also contribute to the global pollution of the air, water, and soil as a result of coal use. In China, for example, where the increasing demands for electricity have been met largely by burning coal, the price in pollution has been high. And China's coal-burning factories have not caused pollution only in China. In a 2004 article in the *Wall Street Journal*, authors Matt Pottinger, Steve Steklow, and John J. Fialka describe how China's increase in burning coal has contributed to the buildup of toxic mercury in the world's food and water supply. They wrote, "Some scientists now say that 30 percent or more of the mercury settling into U.S. ground soil and waterways comes from other countries—in particular, China."

Those who are less enthusiastic about U.S. coal reserves point to the fact that the quality of the mined coal is ebbing and its EROEI is dropping. According to John Gever, Robert Kaufmann, David Skole, and Charles

Vorosmarty, authors of *Beyond Oil*, "coal may cease to serve as a useful energy source in only two or three decades" due to the growing energy needed to mine it and the decreasing energy density it returns. Still, future efforts to produce energy from existing coal supplies may involve the use of improved technologies. For example, factories that remove hydrogen from coal and then use it to power fuel cells are now in development. New technologies in harnessing the energy from coal could improve its efficiency by more than 35 percent.

Nuclear Power

One of the most talked about alternative energy sources is nuclear energy, which is already used around the world to produce electricity. This is especially true in countries such as France, where it powers roughly 77 percent of the nation's total electricity, and Belgium, where 56 percent of electrical power is derived from nuclear generators. In America, only about 20 percent of all electrical power is currently generated by atomic reactors.

Although nuclear power is clean burning and inexpensive, American public opinion of nuclear reactors hasn't improved much since the serious reactor accidents at Three Mile Island in Pennsylvania in 1979 and Chernobyl in the former Soviet Union in 1986. While nuclear power is considered safe, a single accident at one reactor could lead to the deaths and sicknesses of tens of thousands

Nuclear power plants like the one pictured above, in Chooz, France, supply the country with nearly 80 percent of its electricity.

of people. There are currently Russians who are sick from Chernobyl's radioactive fallout. Since the accident, many women have had children with extreme mental and physical deformities. In addition, food and water from certain areas of Russia are no longer safe to eat or drink. In a 2006 report, the environmental agency Greenpeace reported that the full impact of the Chernobyl disaster could "top a quarter of a million cancer cases and nearly 100,000 fatal cancers." Estimated statistics from the International Atomic Energy Agency (IAEA), however, are far lower. It claims that just fifty people have died from the accident and fewer than 50,000 people are either currently stricken or will become stricken with cancer from radioactive fallout. Actual numbers are probably somewhere in between Greenpeace and IAEA estimates. Either way, the dangers from nuclear accidents remain very real in the eyes of most Americans.

Despite concerns about exposure to radioactive fallout, the U.S. government developed the Nuclear Power 2010 program to help sway public opinion and encourage the establishment of new nuclear plants. Citing a 2006 Annual Energy Outlook Report by the Energy Information Administration, experts claim that American demand for electricity will increase at least 45 percent by 2030. They therefore believe that installing new nuclear reactors is vital. To provide for this increase in demand, the Bush administration passed the Energy Policy Act of 2005. This legislation will offer financial support for six new nuclear power plants ($500 million for the first two, and $250 million divided between the remaining four). It will also impose tighter safety restrictions on the new plants, which will be built beginning in 2010. More than twenty other U.S. nuclear reactors are currently in various stages of construction and planning. Ultimately, nuclear power is going to figure largely in the scope of America's future energy production.

Renewable Energy Sources

Harnessing nature's own energies from the sun, ocean, and wind captures renewable sources of energy. In fact, using wind and water currents as energy dates back to the European Middle Ages, when water mills were used to grind grain and ships were powered exclusively by sails.

Wind

Today, wind turbine technology is advancing rapidly, and interest in installing windmills is growing each year. By far the greatest drawback in some people's opinions is that the quality of the landscape is eroded when hundreds of windmills are installed in so-called wind farms. Much of this has to do with the fact that the strongest winds are those coming off of oceans. People are therefore concerned that desirable coastal property will in the future likely include wind farms on both the east and west coasts of the United States. Despite people's reluctance, the potential benefits from harnessing wind power are great. Energy converted from wind is inexpensive to produce, is efficient, and causes no pollution. Some experts believe that wind power could eventually supply up to half of America's electrical energy needs. The Bush administration's Advanced Energy Initiative includes $44 million for research into

This wind farm on the San Gorgonio Mountain Pass in California's San Bernadino Mountains contains more than 4,000 windmills. It provides enough electricity to power Palm Springs and Coachella Valley, California.

improved wind energy efficiency and the production of wind farms on federal lands.

Hydroelectric Power

Hydroelectric power, like wind power, is a renewable energy source that is already in use. Hydroelectric dams can now be found on most major rivers throughout the world. About 9 percent of America's electricity comes from the gravitational pull of dammed waterways. Conveniently, electricity created from hydropower can be stored easily. Because water can be rechanneled and pumped uphill using its surplus power, hydropower also has a significantly high EROEI. The main problem with hydropower is the toll that man-made dams take on the environment. Over time, dams can evaporate streams, interfere with marine biology, and cause natural waterfalls to dry up. Although few, if any, dams are now being built in the United States, there is great promise in the future of Canada's hydroelectric plans. It would then be possible for the United States to import greater percentages of its electrical power directly from Canada.

Solar Energy

Although solar energy is among the oldest renewable sources of energy (Greeks and Romans used mirrors to direct the sun's rays in order to ignite fires), it has been a difficult energy to convert and store. Typically, solar energy is captured in photovoltaic (PV) cells that generate an electrical current. Although similar technologies have been around for the past century, they have not been affordable until roughly the last decade. Still, installing a PV system is very expensive and far beyond the range of most budgets. The greatest expense comes from the batteries that must be installed in order to store power for times when the sun is not shining. To avoid this expense, many homeowners are tying into the commercial power grid and selling (instead of storing) their excess power. In these cases, homeowners are converting their excess power to a dollar amount that they can then apply to their electric bill if needed. When without sunshine, the homeowners can purchase electricity directly from the local electric company because they are tied into the grid. Even so, few private homeowners have taken advantage of solar power because of the initial costs and the maintenance that is required to keep systems running smoothly.

Several key developments in PV technology may introduce solar power to the masses, however. Two inventions, thin film panels and dye coatings, could drastically reduce the costs of operating PV systems for private homes. As part of the Bush administration's Advanced Energy Initiative, developments in PV technology are being researched. It is hoped that one day Americans will build "zero energy" homes that produce more energy than they consume by incorporating PV cells directly into building materials. The 2007 national budget includes a $148 million Solar America Initiative to expand research into PV technologies. The EROEI for solar power will increase as the cost for building solar-powered systems becomes more affordable and universal.

Hydrogen

Hydrogen technology (using liquid hydrogen to store energy) has been discussed as a long-term solution to

America's growing energy demands. Hydrogen, along with carbon, is the basis of all fossil fuels. Although hydrogen atoms bind together with oxygen atoms to create water (H_2O), which is abundant, hydrogen does not exist on the earth in a pure state. Pure hydrogen is man-made and must be produced in one of two ways: by steam reforming or electrolysis.

Both methods of creating pure hydrogen to store energy are possible, but steam reforming gives off carbon emissions in the form of methane. Electrolysis splits the hydrogen atom from water and has no carbon emissions since nothing is burned, but it's a very expensive process. It is this electrolysis that scientists are trying to develop for future hydrogen energy storage. While the process has great potential as a future energy supply, it will take years to create the infrastructure needed to build and establish the facilities needed to create, store, and transport pure hydrogen.

Biomass, Biodiesel, and Ethanol

Many things, including plants, animal waste, and agricultural waste such as corn stalks, wood, and even sugar cane, can be used as a source of renewable power. All of these materials are known as biomass when they are converted into energy. Unlike wind power, solar power, or hydropower, biomass produces pollution, which in some cases is extreme. In India, for instance, where cow

Ethanol, made from corn, is commonly used to supplement gasoline. Above, a factory worker unloads his grain truck at the Adkins Energy plant in Lena, Illinois. An average bushel of corn produces 2.7 gallons (10.2 liters) of ethanol.

dung is burned for cooking fuel, the soils are depleted of necessary nutrients and the air gets clouded with haze. Burning wood also produces toxic substances that cause air pollution. Because of these side effects and also because biomass fuels have extremely variable EROEIs, they will likely not make up much of the energy sources of the future.

When burning oil for transportation, biodiesel is a beneficial alternative to regular gasoline. It is now possible to run a diesel-powered vehicle with what is essentially chemically altered cooking oil (a mixture of vegetable oil, methanol, and lye). Some people even use recycled cooking oil from restaurants for this purpose. Biodiesel also produces fewer harmful air pollutants than traditional gasoline does. Still, biodiesel's EROEI is low, and it is an unlikely answer to the country's transportation fuel needs.

Ethanol use is far more widespread than biodiesel. Ethanol is a type of alcohol that is made from fermentation of grain (like corn stalks or sugar cane), which can be used to supplement gasoline in automobiles. Although ethanol burning is cleaner than gasoline, its EROEI is much less promising. However, when ethanol is blended with gasoline, it decreases carbon emissions. Typical vehicles use a fuel called E10. It is a mix of 10 percent ethanol and 90 percent gasoline. Flexible fuel vehicles (FFV) can use mixtures of up to 85 percent ethanol, or E85. Today, ethanol is used in 46 percent of America's gasoline and uses approximately 20 percent of the nation's corn crop. Its use is expected to increase annually.

For all of ethanol's promise, however, many people still feel that conservation is a better choice when it comes to decreasing our demand for foreign oil imports. In a January 2007 op-ed column in the *New York Times*, Paul Krugman compares the costs of ethanol use and general fuel conservation, writing, "The Congressional Budget Office estimates that reducing gasoline consumption 10 percent through an increase in fuel-economy standards would cost producers and consumers about $3.6 billion a year. Achieving the same result by expanding ethanol production would cost taxpayers at least $10 billion a year, based on the [federal] subsidies ethanol already receives—and probably much more, because expanding production would require [additional] subsidies."

The Importance of Conservation

Just like raising fuel-economy standards will reduce the amount of oil we use, so will conservation. The more we understand how precious oil is, the more we will appreciate it and not waste it. The future's energy strategies are uncertain. It is up to each person to elect leaders who are up to the challenge of meeting tomorrow's energy demands with a plan that goes far beyond oil. Perhaps there will even be the development of an entirely new type of energy that hasn't yet appeared. In the meantime, conservation is key. A sound conservation program, changes in how we use crude oil, and the addition of and investment in renewable energies will help America and the world move forward positively.

Glossary

barrel A unit of volume equal to forty-two U.S. gallons.

biomass Liquid fuels produced from plant feed stocks.

bipartisan Supported by members of two parties, especially in politics.

carbon dioxide A colorless, odorless noncombustible gas that is present in the atmosphere.

commodity Goods such as cattle, sugar, oil, or gold that can be transported and used for commerce.

desalinization The removal of salt from sea water.

dictatorial Of or relating to a dictator, one who rules absolutely and often oppressively.

diesel A low-cost petroleum-based fuel commonly used in trucks.

diplomacy The delicate negotiations between nations.

electrolysis The passage of electricity through an ion-containing solution in order to produce chemical changes in that substance.

embargo An economic boycott put in place by a government or other organized body.

entrepreneur A person who sets up and finances a commercial business or enterprise to make a profit.

EROEI (energy returned on energy invested) The ratio of net energy that is produced after the energy needed to harvest it is subtracted.

fossil fuel Organic matter from plants and animals that died and were compressed over billions of years and turned into oil, petroleum, and natural gas.

fuel cell An electrochemical cell that uses a chemical reaction between hydrogen and oxygen to produce electricity.

global warming The process of gradual heating in the atmosphere caused by the burning of fossil fuels and industrial pollutants.

greenhouse gases Gases that trap the heat of the sun in the earth's atmosphere, producing the greenhouse effect; they include water vapor, carbon dioxide, methane, ozone, chlorofluorocarbons, and nitrous oxide.

Industrial Revolution A period in history when goods became mass-produced by machine in large factories instead of individually by hand.

infrastructure The public facilities of a country or city such as its systems of roads, bridges, dams, sewers, tunnels, and highways.

kickback A slang term referring to money given to a party as part of a secret agreement.

monopoly A company or group of companies' sole control of an industry brought about by elimination of competition.

obtuse Slow to comprehend or perceive.

OPEC (Organization of Petroleum Exporting Countries) An intergovernmental agency created in 1960 by Iran, Iraq, Kuwait, Saudi Arabia, and Venezuela.

photovoltaic Capable of producing electricity when exposed to sunlight.

radioactive fallout Tiny particles of radioactive debris in the atmosphere that slowly fall to the earth following a nuclear explosion. Human exposure to radioactive fallout can cause cancer and death.

refute To prove something false; to discredit.

revenue Another word for capital, income, or money.

smelt To melt or fuse ore in order to separate its metallic components.

subsidy Monetary assistance granted by a government to a person or enterprise.

volatile Characterized by or prone to sudden change.

For More Information

Canadian Energy Research Institute (CERI)
#150, 3512–33 Street NW
Calgary, AB T2L 2A6
Canada
(403) 282-1231
Web site: http://www.ceri.ca

National Energy Board
444 Seventh Avenue SW
Calgary, AB T2P 0X8
Canada
(800) 645-5605 or (403) 292-4800
Web site: http://www.neb.gc.ca

United States Department of Energy
Energy Efficiency and Renewable Energy (EERE)
Mail Stop EE-1
Washington, DC 20585
(800) DIAL-DOE (342-5363)
EERE Information Center: (877) 337-3463
Web Site: http://www.eere.energy.gov

United States Energy Association (USEA)
1300 Pennsylvania Avenue, Suite 550

Washington, DC 20004

(202) 312-1230

Web site: http://www.usea.org

World Energy Efficiency Association (WEEA)

1101 15th Street NW, Suite 1100

Washington, DC 20005

(202) 778-4942

Web site: http://www.weea.org

World Watch Institute

1776 Massachusetts Avenue NW

Washington, DC 20036-1904

(202) 452-1999

Web Site: http://www.worldwatch.org

Web Sites

Due to the changing nature of Internet links, Rosen Publishing has developed an online list of Web sites related to the subject of this book. This site is updated regularly. Please use this link to access the list:

http://www.rosenlinks.com/itn/otef

For Further Reading

Bickerstaff, Linda. *Oil Power of the Future: New Ways of Turning Petroleum into Energy* (Library of Future Energy). New York, NY: Rosen Publishing Group, 2002.

Bily, Cynthia A., ed. *Global Warming* (Opposing Viewpoints). Farmington Hills, MI: Greenhaven Press, 2006.

Gunkel, Darrin, ed. *Alternative Energy Sources* (Current Controversies). Farmington Hills, MI: Greenhaven Press, 2006.

Miller, Kimberly M. *What If We Run Out of Fossil Fuels?* (What If?). Danbury, CT: Children's Press, 2002.

Morris, Neil. *Fossil Fuels* (Energy Sources). Mankata, MN: Smart Apple Media, 2006.

Passero, Barbara, ed. *Energy Alternatives* (Opposing Viewpoints). Farmington Hills, MI: Greenhaven Press, 2006.

Segall, Grant. *John D. Rockefeller: Anointed with Oil* (Oxford Portraits). New York, NY: Oxford University Press, 2001.

Snedden, Robert. *Energy Alternatives* (Essential Energy). 2nd ed. Chicago, IL: Heinemann, 2006.

Snedden, Robert. *Energy from Fossil Fuels* (Essential Energy). 2nd ed. Chicago, IL: Heinemann, 2006.

Bibliography

Gelbspan, Ross. *The Heat Is On*. Updated ed. Cambridge, MA: Perseus Books, 1998.

Gever, John, Robert Kaufmann, David Skole, and Charles Vorosmarty. *Beyond Oil: The Threat to Food and Fuel in the Coming Decades*. Boulder, CO: University Press of Colorado, 1991.

Greenpeace.org. "Chernobyl Death Toll Grossly Underestimated." April 18, 2006. Retrieved January 31, 2007 (http://www.greenpeace.org/international/news/chernobyl-deaths-180406).

Heinberg, Richard. *The Party's Over: Oil, War, and the Fate of Industrial Societies*. Gabriola Island, BC, Canada: New Society Publishers, 2003.

Krugman, Paul. "The Sum of All Ears." *New York Times*, January 27, 2006. Retrieved January 31, 2007 (http://select.nytimes.com/2007/01/29/opinion/29krugman.html).

Kunstler, James Howard. *The Long Emergency: Surviving the End of the Oil Age, Climate Change, and Other Converging Catastrophes of the Twenty-first Century*. New York, NY: Atlantic Monthly Press, 2005.

McKibben, Bill. *The End of Nature*. New York, NY: Random House, 1999.

Office of the Press Secretary. "State of the Union: The Advanced Energy Initiative." Retrieved January 31, 2007 (http://www.whitehouse.gov/news/releases/2006/01/20060131-6.html).

Palast, Greg. "Secret U.S. Plans for Iraq's Oil." BBCNews.com. March 17, 2005. Retrieved February 4, 2007 (http://news.bbc.co.uk/go/pr/fr/-/2hi/programmes/newsnight/4354269.stm).

Pottinger, Matt, Steve Stecklow, and John J. Failka. "Invisible Export—A Hidden Cost of China's Growth: Mercury Migration." *Wall Street Journal*, December 20, 2004.

Ridgeway, James. "Bush Administration Sees New Climate Change Report, Says, 'Whatever.'" *Mother Jones* online. Retrieved January 31, 2007 (http://www.motherjones.com/blue_marble_blog/index.html#3435).

Roberts, Paul. *The End of Oil: On the Edge of a Perilous New World*. New York, NY: Houghton Mifflin, 2005.

Rosenthal, Elisabeth, and Andrew C. Revkin. "Science Panel Calls Global Warming 'Unequivocal.'" *New York Times* online. February 2, 2007. Retrieved February 3, 2007 (http://www.nytimes.com/2007/0203/science/earth/03climate.html).

Shaw, Jonathan. "Fueling Our Future." *Harvard Magazine*. May–June 2006. Retrieved February 1, 2007 (http://www.harvardmagazine.com/on-line/050692.html).

Index

A

Advanced Energy Initiative, 39, 47, 50
alternative energy sources, 15, 27, 39, 41–54
Arab oil embargo of 1973–74, 16, 26

B

Beyond Oil, 44
biodiesel, 52–53
biomass, 51–52
Bush, George W., and oil, 15, 17, 32, 35, 39, 46

C

Chernobyl disaster, 44–45
coal use, 18–20, 39, 41, 42–44
commodity, oil as a, 6–7
conservation, 15, 31, 36–38, 41, 54
crude oil, 6, 9, 10, 12, 15, 17, 27, 41

E

electrolysis, 51
The End of Nature, 43
The End of Oil, 30
Energy Information Administration (EIA), 9, 31, 46
Energy Policy Act of 2005, 46
EROEI, 28, 40, 42, 43, 49, 50, 52, 53
ethanol, 53–54

F

First Gulf War, 12, 14
fuel cells, 44

G

gasoline, as an oil by-product, 4
geopolitics, 11–13
global warming, 33–35
greenhouse gases, 39
Greenpeace, 45, 46

H

Hurricane Katrina and oil, 9–10, 17
hydroelectric power, 41, 44, 50–51

I

Intergovernmental Panel on Climate Change (IPCC), 33, 35
International Atomic Energy Agency (IAEA), 45, 46

K

kerosene use, 21, 23
Kyoto Protocol, 35

M

Middle East and oil, 6, 9, 11–13, 15–17, 23, 26, 31, 39, 42

N

natural gas, 41–42
9/11 terrorist attacks, 14, 15
nuclear power, 27, 44–46

O

OPEC, 13–14, 15, 26, 31

P

peak oil, 28–32, 40
petroleum, defined, 5–6
photovoltaic (PV) cells, 49, 50

R

Rockefeller, John D., 21–23

S

Sherman Anti-Trust Act, 24
solar power, 39, 41, 49–50, 51
Strategic Petroleum Reserve
(SPR), 15–17
SUVs, 32

U

U.S. Energy Information
Administration (EIA), 41, 42
U.S. oil, history of, 6, 16, 20–27

W

wind power, 39, 41, 47–48, 51

About the Author

Joann Jovinelly is an editor and award-winning author. She has written many books for young adults on various subjects ranging from ancient and medieval history to politics and current events. Jovinelly lives in New York City.

Photo Credits

Cover, back cover, pp. 4, 8, 16, 18, 22, 25, 33, 39, 52 © Getty Images; p. 5 © Martin Shields/Photo Researchers, Inc.; p. 10 © AP Images/Peter Cosgrove; pp. 13, 28, 34, 37, 45, 48 © AFP/Getty Images; p. 19 Library of Congress Prints and Photographs Division; p. 27 © Larry Lee Photography/Corbis; p. 29 Courtesy M. King Hubbert Papers, American Heritage Center, University of Wyoming; p. 32 © Joanna B. Pinneo/Aurora/Getty Images.

Designer: Tom Forget; **Photo Researcher:** Cindy Reiman